Basic Biology:
Homeschooling Week By Week

John Turano

Publisher: John Turano
ISBN-13: 978-1466342170
ISBN-10: 146634217X

Cover Design: Holly Lisle

Cover Art:
Copyright © Sascha Burkard / BigStock.com

3/20/23 assignt

INTRODUCTION

I love biology. I have for a long time and teaching it has been one of the joys of my professional life. Having the opportunity to pass on the discovery of the natural world is something I have never gotten tired of. Every day, it seems, there is some new discovery, or new explanation of a natural phenomenon, or a re-interpretation of established beliefs.

Unfortunately, public education has gone test-score crazy and science curricula, along with all the other subjects, has shifted to a teach-to-the-test mentality. It's the test that drives what the kids are supposed to know. How wrong is that?

You, as a homeschool parent, don't have those restrictions. You're only restricted by what you think is important for your child to learn to make him or her a better person later in life. This guide is for you.

Teaching your homeschool child elementary level science isn't as difficult as teaching a middle-school level child and it's a lot less difficult than teaching a high-school level child. In this guide I'm going to provide you with a framework for a 36-week basic high-school biology course. At the beginning of every unit I've given you a timeline for the modules in the unit. These are only guidelines based on my own experience teaching these topics over the years. The timeline is not etched in cement. My intent is to give you as much flexibility as possible to go into each topic in as much depth as you think is appropriate for your child.

But beyond that, I'm providing you, in this guide, with a single focal point. A point that is not brought out in most basic biology courses or textbooks.

Let me explain. At the beginning of every year I go through an exercise with my kids in defining what makes a living thing, a living thing. What characteristics do all living things share? It's a standard exercise to help get the kids thinking about the differences between living and non-living things. The real obvious ones like eat and breathe come up quickly. I'll get references to reproduction and movement. But when I ask them why an organism needs to breathe or eat I get their logic loop.

Here's what I mean. Ask the kids why an organism needs to eat, they'll correctly say, "In order to stay alive." Right. "But why does an organism need to stay alive?" I'll ask. And here's where I'll either get the logic loop of in order to eat, or as often times happens, I get silence.

You see, biology isn't taught to ask the why of living things. It's not taught to ask the question of why organisms do what they do, the single driving force that every living thing has in common, if you will. We dance around the question of why but never really examine it. It's not on the test, you see.

What always bothered me to a certain degree was that we could teach biology from different themes. You could teach it from an ecological viewpoint, or a molecular viewpoint, or even a taxonomic viewpoint, but never from a single unifying viewpoint. I always wondered why we didn't do this, since both chemistry and physics have a singular theme that weaves throughout both fields.

And then I had my "Ah-ha!" moment. I read a book called *The Selfish Gene,* by Richard Dawkins, in which he explained the whole idea of why we have genes. I hope he'll forgive me for over-simplifying this magnificent work but it made so much sense to me. The pieces then fell into place.

An organism's genetic code exists to ensure that the species will survive in successive generations. In order to do that, the organism must reproduce successfully, and in sufficient numbers. According to Dawkins, the organism's "body", in whatever form that may take, is nothing more than a vessel to carry the genetic code. That's the selfish part. All organisms must reproduce successfully so that the species survives in order that the genetic code for that species is protected generation after generation.

In other words, everything an organism does is done to prevent extinction of the species. Protect the genetic code at all costs!

I developed this home study guide to basic biology with that single premise in mind. How do organisms prevent their own extinction? What activities must they perform to ensure successful reproduction? Now when you examine characteristics of living things, it makes more sense. Of course all organisms must perform the same functions. They're all after the same thing, the prevention of their own extinction!

I've taken eighteen topics and separated them into four units. I also designed the units to cover a 36-week course of study. As you'll notice, the units have a common theme to them. This guide will allow you a great deal of flexibility to go into any topic in as much depth as you feel is appropriate.

UNIT 1 – How Life is Structured
UNIT 2 – How Life Replicates
UNIT 3 – How Life Adapts
UNIT 4 – How Life Interacts

Each Unit consists of a number of modules with Unit 1 being the largest, at six modules. The modules are all laid out the same. There is a brief overview of the module, followed by the goals for you to meet in the module, then key words (vocabulary is King in biology!), and finally an activity that you can do.

The activities I've chosen are simple and, for the most part, inexpensive. I've included those that utilize items that you may already have in your house. And even though the measurements are metric, most measuring cups will have both English and metric measure on them.

I've tried to make the units free standing, with the exception of the first unit. There has to be a common starting point, and when you examine the topics in the first unit you'll understand why. Beyond that, you can move the units around as you see fit.

My goal is to provide you with a syllabus that will allow you to teach your child about biology in a manner that is appropriate to his or her life, that will have a lasting effect on how your child sees the living world, and to create a deeper understanding of how wonderful the living world truly is.

Just remember this simple idea: Biology is the study of living things and how they prevent their own extinction.

Unit 1 - How Life Is Structured

This unit consists of six modules and is designed to lay the foundation for the other units and modules.

Time Frame: 9 – 10 weeks

Module 1 Science, Biology, and Living Things: 2 – 3 weeks
Module 2 Biochemistry: 2 weeks
Module 3 Cell Structure and Function: 1 week
Module 4 Photosynthesis: 1 week
Module 5 Cellular Respiration: 1 week
Module 6 Cell Growth and Division: 2 weeks

end 3/20/23

John Turano

Module 1: Science, Biology, and Living Things

Module 1 Overview

This module is an overview of science, of what it is and what it is not. Scientific discovery is built on the work of others who have lived before. It's important to the study of biology, and science in general, that those discoveries be put in historical context. On that point you can spend as much or as little time as you want. The history of science is filled with intrigue and drama.

Of course you need to study how science and scientists work. Time needs to be spent on the method of scientific inquiry and fully understand the gross misconceptions of modern science—misconceptions about the term "theory" for example. It's critical that these misconceptions are cleared up and understood in the context of scientific terminology.

A good example of this is Einstein's Theory of Relativity or Darwin's Theory of Natural Selection. I also like to use the centuries-old belief in spontaneous generation and how it was finally disproved as an example of how some discoveries that go against conventional wisdom take a long time to be accepted.

Scientific measurements in the form of the International System (SI), or its common name—the metric system—can also be done at this point. You can do scientific notation, but you rarely see that in biology, along with graphing, which is widely used in biology.

If the goal of every living thing is to prevent extinction of the species, and if you keep that in mind throughout this entire course, then you have to come up with a definition of what a living thing is and what separates it from a non-living thing.

I always like the *Star Trek* analogy of traveling to another part of the galaxy and coming up on an alien life form. How are we going to tell if it's alive by our definition? What criteria are we going to take with us on these travels into the far corners of the galaxy to determine if there is life out there?

Now you're ready to talk about the characteristics of living things! And no matter how simple or complex living things are, they all share the same life characteristics.

3/21/23
Read

Module 1 Goals

The following is a list of goals that should be met by the end of this module. I've made note of those that are critical, important, or nice-to-know. The critical ones can't be missed. The important ones can be missed but will only add depth if you want further study. The nice-to-know ones are exactly that. If you have the time, or your really want some depth, cover those as well.

Explain what the goal of science is. (**Critical**)
Differentiate between a theory and a hypothesis. (**Critical**)
Explain what a hypothesis is. (**Critical**)
Describe how scientists test hypotheses. (**Critical**)
Explain how a scientific theory develops. (**Critical**)
Describe the characteristics of living things. (**Critical**)
Explain how life can be studied at different levels. (**Critical**)
Define biology and describe the various branches of biology. (Important)
Describe the measurement system most scientists use. (Important)
Explain how light microscopes and electron microscopes are similar and different. (Nice-to-know)

Module 1 Key Words

Biology is a science of words. Understanding the definitions of key terms goes a long way to understanding the science. Here's a list of key words for this module:

adaptation
biodiversity
biology
biosphere
biotechnology
cell
constant
data
dependent variable
ecosystem
evolution
experiment
gene
genomics
gram
homeostasis
hypothesis
independent variable
liter
metabolism
meter
microscope
molecular genetics
observation
organism
species
system
theory

[handwritten annotations] Vocab for 3/22/23. Use index cards so you can study them. Read/write to understand these.

A. 3/22 Vocab

Index Cards Read/write to Understand

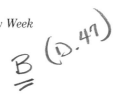

Module 1 Activity B (p.47)

PAPER FOLDING - A HYPOTHESIS

Here's a real neat activity that's easy to do and will demonstrate very nicely the principle behind forming a hypothesis along with the principles involved in experimental design.

OBSERVATION

All paper (writing paper, copy paper, paper napkins, toilet paper, construction paper, newspaper) is made out of the same material, wood. Each of these different types of paper have different thicknesses. Because of the different thicknesses, can they all be folded "x" number of times. The question is, how many times can a piece of paper be folded?

MATERIAL

In order to do this activity somewhat scientifically, you need samples of as many different types of paper that you can find. When I do this I use things like a piece of paper towel or paper napkin, construction paper, printer paper, lens or filter paper and toilet paper. You could also use newspaper, paper bags, or loose leaf paper. And if you're really brave, a piece of a corrugated box! The goal here is to get at least 4 or 5 different types of paper.

For this to work well you need five samples of each type of paper you selected.

HYPOTHESIS

A hypothesis is a prediction of an outcome based on prior knowledge. So before you make a hypothesis of how many times you can fold a piece of paper, examine your samples and then decide how many times you can fold that sample.

To keep track of your hypotheses you can construct a simple

chart, like the one below, to record your predictions. In nay case, your hypothesis is a statement that resembles the following:

A piece of _____ paper can be folded "x" times.

TYPE OF PAPER	# OF FOLDS

PROCEDURE

The procedure is very straightforward. You're going to fold each sample as many times as you can and record the number of folds in a table that looks like the one below.

DATA TABLE

TYPE OF PAPER	# Folds	# Folds	# Folds	# Folds	# Folds	AVG

DATA ANALYSIS

Every well-designed experiment must include an analysis of the data. This is done to support the hypothesis. If the data doesn't support the hypothesis, something is wrong and the errors need to be identified and addressed, usually in another series of experiments.

In this activity there are variables that have not been accounted for. This is intentional since part of the purpose here is to lead the student to finding these variables and address the possible errors these variables create.

- What is the average number of folds for each type of paper? What is the average number of folds for all samples?
- What factors influenced the total number of folds for each sample of paper?
- What variables were not taken into consideration with this experiment?

Repeat the experiment taking these variables into consideration and record the results again. Were the new set of resulting data different?

CONCLUSIONS
After all the variables have been accounted for and all the data analyzed, a conclusion can be reached.

Based on the above data analysis, what conclusions can you make about how many times a piece of paper can be folded?

D
Quiz - VOCAB. 10pts.

E. Quiz over Module 1.
Quiz will cover Module 1
GOAls as on page 11. 10pts

Module 2: Biochemistry

DAy 1
Module 2

Module 2 Overview   *Read*

This module provides the background needed for understanding the chemical nature of biological organic compounds. As a result of this, a common chemical thread begins to appear and, hopefully, you see that all living things on Earth are made out of the same "stuff".

So it doesn't matter how big or small an organism is, or how simple or complex it is, a carbohydrate is a carbohydrate. And the same is true for the other classes of organic compounds.

You can start out with a quick review of what matter is and the forms that matter exists in. The Law of Conservation of Mass and Energy is good to reinforce the idea that all the material we need is already here. Of course when you talk about matter you need to go into the structure of atoms. This opens the door for understanding chemical bonding.

Before you do chemical bonding you need to examine how the electrons in an atom are arranged. I like to keep this as simple as possible, much to the dismay of my chemistry teacher friends, I might add. Chemical bonding can be a tough concept to understand so I break it down as simply as I can. But understanding electron levels, or energy levels as they are sometimes known, is important.

At this point you could skip over to the Periodic Table of Elements. I explain how the Periodic Table is organized and let my chemistry teacher friends worry about the rest! I believe that if the kids know how the Periodic Table is organized, understanding why sodium bonds to things like chlorine or fluoride will make more sense. And once they see the pattern in the columns and rows it's like an "A-ha" moment and that's a beautiful thing.

Now you're ready to explain the three types of chemical bonds that exist. If the idea of energy levels is understood, the explanation of the different types of bonds goes a lot easier.

From here you go into a description of what organic chemistry is all about and why the element carbon is so important. All life on Earth is built around the carbon atom and there's a reason for this. In fact, carbon is so ideal that many scientists believe that if life is found "out there," it's very probable that it will also be based on carbon. That's way cool!

Now you need to describe the characteristics of the four classes of organic compounds found in all living things: carbohydrates, proteins, fats, and nucleic acids. You also need to consider a special category of protein called enzymes. And of course, you need to spend time on water. Everybody knows that water is important for life on the planet but how many understand why? Consider this about water, for example: The giant Sequoia trees of California can grow to 300 feet or so. How does water defy gravity and get way to the top of those humongous trees?

Once you've completed this module, you have the basics necessary for later topics like photosynthesis, cellular respiration, and how the DNA molecule is held together.

Day 2
Mod. 2

Module 2 Goals

The following is a list of goals that should be met by the end of this module. I've made note of those that are critical, important, or nice-to-know. The critical ones can't be missed. The important ones can be missed but will only add depth if you want further study. The nice-to-know ones are exactly that. If you have the time, or your really want some depth, cover those as well.

Identify the three subatomic particles found in atoms. **(Critical)**
Explain what chemical compounds are. **(Critical)**
Describe the three main types of chemical bonds. **(Critical)**
Explain why water molecules are polar. **(Critical)**
Describe the function of each group of organic compounds. **(Critical)**
Explain how chemical reactions affect chemical bonds in compounds. **(Critical)**
Explain why enzymes are important to living things. **(Critical)**
Differentiate between solutions and suspensions. (Important)
Describe how energy changes affect how easily a chemical reaction will occur. (Important)
Explain how all of the isotopes of an element are similar and how they are different. (Nice-to-know)
Explain what acidic solutions and basic solutions are. (Nice-to-know)

DAy 2 Mod 2
John Turano

Module 2 Key Words

Here's a list of key words for this module:
acid
activation energy
adhesion
amino acid
atom
base
bond energy
carbohydrate
catalyst
chemical reaction
cohesion
compound
covalent bond
element
energy level
enzyme
fatty acid   DAy 3
hydrogen bond   Mod 2
ionic bond
lipid
molecule
nucleic acid
pH
polymer
product
protein
reactant
solute
solution
solvent
substrate

DAY 4
Mod 2.

Module 2 Activity

RED CABBAGE JUICE pH INDICATOR

In the laboratory, pH paper and chemicals are commonly used to indicate pH. In the homeschool, students can make their own pH indicator using red cabbage juice, which changes color in the presence of an acid or base. The plant pigment anthocyanin is the active ingredient responsible for the color change. In this activity, students make a pH indicator from red cabbage juice and then use it to test various substances.

MATERIAL
- red cabbage leaves
- 200-mL water (about 6.5 oz)
- small strainer
- clear container (250mL/1-cup)
- medicine droppers (1 per test substance)
- small clear containers or clear drinking cups (1 per test substance)
- 1-mL of each of the following:
- vinegar, household ammonia, lemon juice, fruit juice, milk, detergent, soda, antacid tablets, baking soda

PROCEDURE
1. Chop up several red cabbage leaves.
2. Place the leaves in a blender and add 200mL (about 6.5 oz) water.
3. Blend the mixture and then pour it through the strainer into the large beaker or container. The red cabbage juice indicator is now ready for use. Note: If you use distilled water the indicator will have a reddish purple color and if you use tap water the indicator will have a violet blue color.
4. Pour 10-mL of the cabbage indicator into two of the small clear containers.
5. Add a few drops of vinegar to one container and a few

drops of ammonia to another. Vinegar is an acid and ammonia is a base. Vinegar turns the indicator red and ammonia turns it green. Use these 2 samples as references for the other test substances.

6. Pour 10mL of indicator into as many containers as you have test solutions. Test each substance by adding a few drops of it to a container the indicator.

7. To test an antacid tablet or other solid, crush it, dissolve it in water, and add a few drops of the resulting solution to a container of indicator.

Module 3: Cell Structure and Function

Module 3 Overview

Now that you have the chemistry module behind you, you can examine the building block of all living things on Earth...the cell.

There is a little history that can be viewed to set the stage for the modern concept of the cell. Of course the discovery of the cell by Robert Hooke is a good starting point. But also is the development of the Cell Theory in the 19th century and the scientists responsible for that. The Cell Theory is a key point in biology since it led to the development of the Germ Theory of Disease.

I usually start off by distinguishing between the two types of cell. There are very distinct characteristics that separate the two categories and understanding what those differences are is important. A thorough explanation of the how's and why's of the differences is best covered in the module on the history of life. Regardless of what type of cell you study, they must all perform the same life functions outlined in the first module.

Most of this module centers on the eukaryotic cell structures. Every one of the cell organelles performs a specific function for the cell in order for the cell to remain alive. Analogies between most cell organelles and human body systems can be made. Distinction has to made between those structures that are found in animal cells and those that are found in plant cells.

Special attention needs to be made on the cell membrane. This amazing structure keeps things inside the cell, keeps things out of the cell, and will allow only certain substances to pass through it. This is a good time to differentiate between osmosis and diffusion. You also need to examine active and passive transport as mechanisms for the movement of substances across the cell membrane.

The last thing that should be covered is how cells organize to form tissues, organs, and systems.

Module 3 Goals

The following is a list of goals that should be met by the end of this module. I've made note of those that are critical, important, or nice-to-know. The critical ones can't be missed. The important ones can be missed but will only add depth if you want further study. The nice-to-know ones are exactly that. If you have the time, or your really want some depth, cover those as well.

Distinguish between eukaryotes and prokaryotes. (**Critical**)
Describe the function of the cell nucleus. (**Critical**)
Describe the functions of the major cell organelles. (**Critical**)
Identify the main roles of the cytoskeleton. (**Critical**)
Identify the main functions of the cell membrane and the cell wall. (**Critical**)
Describe what happens during diffusion. (**Critical**)
Explain the processes of osmosis, facilitated diffusion, and active transport. (**Critical**)
Describe cell specialization. (Important)
Explain what the cell theory is. (Important)
Describe how researchers explore the living cell. (Nice-to-Know)
Identify the organization levels in multi-cellular organisms. (Nice-to-Know)

Module 3 Key Words

Here's a list of key words for this module:
active transport
cell membrane
cell theory
cell wall
centriole
chloroplast
concentration gradient
cytoplasm
cytoskeleton
diffusion
endocytosis
endoplasmic reticulum
eukaryotic cell
exocytosis
facilitated diffusion
fluid mosaic model
Golgi apparatus
hypertonic
hypotonic
isotonic
lysosome
mitochondrion
nucleus
organelle
osmosis
passive transport
phagocytosis
phospholipid
prokaryotic cell
receptor
ribosome
selective permeability
vacuole
vesicle

Module 3 Activity

MODELING THE CELL

The diversity of life on Earth is enormous, although all living things are made from the same basic structural unit, the cell. In your body alone, there are trillions of cells. In this activity, you will make a model of a cell. And if it doesn't work out well, you have some jello to eat!

MATERIALS
- 2 resealable plastic sandwich bags
- Jell-O jigglers
- a small round balloon
- a permanent marker
- coffee stirring straws
- drinking straws cut in half
- different sizes of erasers
- slices of two colors of sponges
- tiny beads

PROCEDURE
1. Use the materials to construct a detailed model of a cell.
2. Be sure to include at least the following components in your model: cell membrane, cytoplasm, nucleus, cytoskeleton, ribosomes, mitochondria, Golgi apparatus, and centrioles.
3. Use both sandwich bags in constructing your model.
4. Tightly seal your cell after it has been completed.

If you're not sure how to make Jell-O jigglers, here's a recipe.
- 2-1/2 cups boiling water or apple juice (Do not add cold water)
- 2 (8-serving) pkg. Gelatin Dessert, any flavor; OR
- 4 (4-serving) pkg. Gelatin Dessert, any flavor

Stir boiling water into gelatin in large bowl at least 3 minutes until completely dissolved.

- Pour into 13x9-inch pan.
- Refrigerate at least 3 hours or until firm (does not stick to finger when touched).
- Dip bottom of pan in warm water about 15 seconds.
- Cut into decorative shapes with cookie cutters all the way through gelatin or cut into 1-inch squares.
- Lift from pan.

Module 4: Photosynthesis

Module 4 Overview

All living things need a source of energy. Of course the primary source of energy is the Sun, but living things can't use solar energy directly. The next two modules deal with energy. You first have to examine how energy is converted from the sun, stored by cells, and then utilized by the cells. In biology the two processes of making, storing energy and utilizing that energy are photosynthesis and cellular respiration. This module is about photosynthesis.

Ironically enough, photosynthesis did not come before cellular respiration. Nor did both processes develop at the same time. Life had been present on Earth for millions of years before cells adapted to converting the Sun's solar energy into chemical energy stored in glucose. Prior to that, life got on very well with cellular respiration only, a process that still goes on today with a number of species.

Photosynthesis also changed everything on Earth. Over a period of millions of years our atmosphere gradually changed and no longer was life confined to the seas. Nor was life restricted to one or two cells. Now life could evolve into land forms and become multicellular. All because of the highly unstable and very poisonous by-product of photosynthesis...oxygen.

This module, and the one on cellular respiration, is chemistry. Understanding the two major reactions of photosynthesis requires an understanding of bonding and how energy is stored in bonds. In addition, understanding how enzymes work is also important.

You can make this module as detailed as you want, but a good understanding of what photosynthesis is all about has to include knowledge of what happens inside the plant cell with the water and carbon dioxide that plant cells use to make glucose. You also need to examine what goes on during the daylight hours and at night when there is no sunlight.

You need to examine the structure of a green leaf and the structure of the chloroplast, where the chlorophyll is located. In addition to that you need to understand why chloroplasts have their own DNA and where the two types of chlorophyll are located. Each type of chlorophyll performs a specific function in photosynthesis. For added depth, investigate why plants are green. I like to grow plants in different colored light to demonstrate this. There is a reason for the green. For some real depth, an examination of the Calvin cycle would give you more information than you'll probably ever need.

Module 4 Goals

The following is a list of goals that should be met by the end of this module. I've made note of those that are critical, important, or nice-to-know. The critical ones can't be missed. The important ones can be missed but will only add depth if you want further study. The nice-to-know ones are exactly that. If you have the time, or your really want some depth, cover those as well.

Describe the role of ATP in cellular activity. (**Critical**)
State the overall equation for photosynthesis. (**Critical**)
Describe the role of light and chlorophyll in photosynthesis. (**Critical**)
Describe the structure and function of the chloroplast. (**Critical**)
Describe what happens in the light-dependent reactions. (**Critical**)
Explain where plants get the energy they need to produce food. (Important)
Explain what the Calvin cycle is. (Important)
Identify the factors that affect the rate at which photosynthesis occurs. (Important)
Explain what the experiments of van Helmont, Priestly, and Ingenhousz reveal about how plants grow. (Nice-to-Know)

Module 4 Key Words

Here's a list of key words for this module:

adenosine triphosphate
ATP synthase
autotroph
Calvin cycle
chlorophyll
heterotroph
light-dependent reactions
NADP+
photosynthesis
photosystem
pigment
stroma
thylakoid

Module 4 Activity

SPINACH CHROMATOGRAPHY

BACKGROUND

All life on Earth depends on the process of photosynthesis. Photosynthesis is a process that plants, cyanobacteria, and algae use to convert CO_2 and water, in the presence of light energy, into sugar. Photosynthesis requires the presence of special pigments that can absorb the energy of light. A pigment is a substance that absorbs light of a particular wavelength. Its color depends upon the color of light that if reflects. For example, a green substance appears green because it reflects green light and absorbs all other colors of light, especially red and blue.

The most important plant pigments in photosynthesis are the "chlorophylls." Green plants contain both chlorophyll a and chlorophyll b. In addition to chlorophyll, the leaves of many green plants also contain one or more other pigments, including "carotenes," which are orange, and "xanthophylls," which are yellow. The presence of these other pigments is masked by the abundance of chlorophyll during most of the year.

The pigments in plant cells can be separated from one another by a technique known as chromatography. Chromatography is a technique for separating and identifying substances in a mixture, based upon their solubility in a solvent. It is one of the most valuable techniques chemists and biochemists use to determine ingredients that give flavor or scent, analyze environmental pollutants, identify drugs in urine, and even separate proteins that can identify evolutionary relationships.

The name chromatography is derived from the Greek words "chroma" and "graph", which mean "color writing". Chromatography was invented in 1910 by a Russian botanist,

Mikhail Tswett, who used it to separate plant pigments. When a dye mixture is placed on a strip of chromatography paper and placed into a solvent solution, the individual substances in the mixture will migrate up the chromatography paper at different rates. The rate of migration is based upon the adsorption capacity of the chromatography paper and the solubility of the sample in the solvent. As the solvent moves up the chromatography paper strip substances in the mixture that are soluble in the solvent are carried along up the strip. On the other hand, the substances that are more attracted to the chromatography paper than to the solvent stop moving and form bands or spots along the paper strip.

THEORY OF PAPER CHROMATOGRAPHY
A small sample of a mixture is placed on porous paper, which is in contact with a solvent. The solvent moves through the paper due to capillary action and dissolves the mixture spot. The components of the sample start to move along the paper at the same rate as the solvent.

Components of the mixture with a stronger attraction to the paper (more polar) than to the solvent will move more slowly that the components with a strong attraction to the solvent (less polar). The difference in the rates with which the components travel along the paper, over time, leads to their separation.

Particular mixtures will have chromatographic patterns that are consistent and reproducible as long as the paper, solvent, and time are constant. This makes paper chromatography a qualitative method for identifying some of the components in a mixture.

MATERIALS
• chromatography paper/coffee filter
• fresh spinach leaf
• metal washer or a quarter

- 6" metric ruler
- pencil
- 50-mL beaker (or a small glass container like a baby food jar)
- 10-mL ethyl alcohol or nail polish remover
- large paperclip

PROCEDURE
1. Using the pencil, draw a fine line across the paper 1-cm from one end of the chromatography paper.
2. At the other end of the chromatography paper, draw a line across the paper 1.2-cm from the end.
3. Place the paper strip on top of a spinach leaf so that the 1-cm line is totally on top of the spinach.
4. Take the metal washer and roll the edge of the washer along the pencil line on the paper strip. Go back and forth a few times and be careful not to rip through the paper strip.
5. Take the open paper clip and carefully puncture a hole in the middle of the line that is 1.2-cm from the other end of the paper.
6. Lower the paper into the beaker slowly. The paperclip will rest across the top of the beaker. Do not let the alcohol to come in direct contact with the chlorophyll on the paper. If you measured the line properly, this will not be a concern.
7. As soon as you have done this, record your start time. You are going to let this run for 15-minutes.
8. Observe what happens to the chlorophyll from the spinach leaf.
9. At the end of the 15-minute time period, record your time and take the paper strip out of the beaker.
10. Remove the paperclip and set it aside.
11. Set the chromatography paper flat on the desk top and give it a minute or two to dry somewhat.
12. Measure the distance, in centimeters, from the pencil line to the first pigment. Record your measurement.

13. Measure and record the distances from the pencil line to the other pigments you have on your chromatography paper.
14. Identify the following pigments:
- Carotenes – faint yellow
- Xanthophylls – yellow
- Chlorophyll A – bright green
- Chlorophyll B – yellow-green
- Anthocyanin – red

ANALYSIS & CONCLUSIONS:

Distance from line (cm)	Color	Pigment Name

1. How many pigments were found in the spinach leaf?
2. Explain why any of the five pigments did not show up.
3. Which of these pigments would be considered the most polar?
4. Which one the least?
5. Many trees have leaves that are green in the summer and red, yellow, or orange in autumn. Where were these colors during the summer? How can they suddenly appear in autumn?
6. In addition to separating plant pigments, what are some other possible applications for paper chromatography.

Module 5: Cellular Respiration

Module 5 Overview

Like the module on photosynthesis, this one is also about energy. In this case, it's about how energy is released by cells.

Cellular respiration in itself is not hard to understand. It's the opposite process of photosynthesis. Instead of making glucose, cellular respiration breaks glucose down to release the energy stored in the bonds of the molecule. What is difficult is tracing the chemical pathways that this process follows.

There are two things to consider. First, you have to consider what happens when cellular respiration takes place without oxygen being present. What is known as anaerobic respiration. We also know this as alcoholic fermentation and we can thank Louis Pasteur for figuring it out so high quality wine and beer could be produced. This form of cellular respiration is also the oldest method of producing energy.

In the time before photosynthesis, one-celled organisms relied on producing energy in the absence of oxygen. The reason for that was simple enough. There wasn't any oxygen to use. Once photosynthesis evolved, organisms had to do one of two things. They had to adapt and find a way to incorporate a highly unstable gas into their metabolism or they didn't adapt and therefore went extinct. Most organisms went extinct.

So the second process to consider is cellular respiration in the presence of oxygen. What we call aerobic respiration. Since this form of respiration is the result of an evolutionary process, it's much more complicated.

What you're interested in is how much net energy come out of the process of cellular respiration, and you'll be surprised at how much that really is.

You need to start with an examination of the cell's

mitochondrion, since this is where all the action takes place. As with the chloroplast, look into why the mitochondrion has its own DNA. Interesting story there.

From the mitochondrion's structure, start with glycolysis, move into fermentation, and then the Kreb's cycle, or as it was once known, the Citric Acid Cycle. You can make this as easy or as challenging as you want by including the enzymes. It's important to include the electron transport chain in either case.

As a great extension to this module, you can consider muscle physiology and investigate why aerobic exercise works so well. You can consider why breathing properly during any aerobic-type activity will prevent the dreaded muscle tightness from lactic acid build-up. It's all tied to how our muscle cells utilize the energy they store.

Module 5 Goals

The following is a list of goals that should be met by the end of this module. I've made note of those that are critical, important, or nice-to-know. The critical ones can't be missed. The important ones can be missed but will only add depth if you want further study. The nice-to-know ones are exactly that. If you have the time, or your really want some depth, cover those as well.

Explain what cellular respiration is. (**Critical**)
Describe glycolysis. (**Critical**)
Name the types of fermentation. (**Critical**)
Compare photosynthesis and cellular respiration. (**Critical**)
Describe the Krebs cycle. (Important)
Identify three pathways the body uses to release energy during exercise. (Important)

Module 5 Key Words

Here's a list of key words for this module:

aerobic
anaerobic
calorie
cellular respiration
electron transport chain
fermentation
glycolysis
Krebs cycle
mitochondrion
NAD+

Module 5 Activity

CELLULAR RESPIRATION IN YEAST

This activity will study yeast to see the result of cellular respiration.

<u>MATERIALS</u>
- Packages of dry powdered yeast. (Regular yeast, not quick rising yeast.)
- Warm tap water or heated water (about 1/4 cup)
- Flour (about 1/2 cup)
- Sugar (about 1 teaspoon)
- Plastic zip sandwich bag
- Empty plastic individual sized drink bottle (empty soda bottle or water bottle). No lid needed.
- Balloon
- Beaker or bowl in which to mix ingredients
- Plastic spoon

<u>PROCEDURE</u>
1. Pour half a package of dry yeast into a bowl or beaker.
2. Examine the dry yeast with a magnifier or microscope.
3. Touch or smell it, and write down a description of its properties. Is it biotic or abiotic? Living or non-living?
4. Add 1/4 cup warm water (warm tap water is OK, but not too hot) to the yeast.
5. Watch what happens to the yeast.
6. Stir with the spoon to dissolve the yeast in the water.
7. Add about a teaspoon of sugar to the yeast in warm water, and stir with the spoon.
8. Pour half of the yeast mixture into the empty plastic drink bottle, and put the balloon over the mouth of the bottle, making sure that all air is pressed out of the balloon first, and that it is flattened out.
9. Let the bottles sit in a warm place or hold them against your body for warmth.

10. Pour about a half cup of flour into a plastic zip sandwich bag.
11. Pour the remaining other half of the yeast mixture into the bag of flour.
12. Seal the zip bag, mashing out all of the air you possibly can.
13. Rub and mash the plastic bag in your hands to thoroughly mix the yeast solution with the flour.
14. Make sure that there is no air in the bag, and then mash all of the dough mixture down into one corner of the bag, and twist the bag around it to keep the dough trapped down in one corner.
15. Hold the plastic bag in a warm place, such as under your armpit, on on your stomach against your skin under your shirt.
16. Or put it under a warm light bulb (but not too hot to keep your hand under).
17. Keep it warm for at least 15 minutes, and then take a look at it.
18. After about 15 minutes, observe the bottle with the balloon AND the bag with yeast and flour.
19. How has the yeast mixture changed in the bottle?
20. How has the balloon changed?
21. How has the yeast and flour mixture changed in the plastic bag? Write down your observations.
22. Keep it warm and check it again in another 15 minutes.
23. Make your observations again.
24. Continue to check the two experiments occasionally during the class, noting any changes.

CONCLUSIONS

A. What is the gas that has formed?
B. What other experiments might you like to try with yeast as an extension to this activity?
C. What environmental conditions might you vary?
D. What else might you add to the yeast?
E. What would happen if you put the yeast and flour

mixture into an oven?

F. What do you predict would happen? What will you end up with?

G. What will happen to the yeast and water in the bottle if you leave it long enough?

H. What will you end up with? (List two possibilities.)

I. What are three products that people need yeast to make for them?

J. How do you think people discovered that yeast would do this?

K. Can you make bread without yeast? What is sourdough bread?

L. What is needed to turn alcohol into vinegar?

M. What organisms are involved?

Module 6: Cell Growth and Division

Module 6 Overview

This last module of Unit 1 addresses the cell cycle and how cells divide. It sets the stage for Unit 2, How Life Replicates.

All cells have a cycle. You need to identify what the four stages of that cycle is and what goes on in each of the stages. This activity centers on the nucleus of the cell and the chromosomes found inside. You can study the structure of the chromosome at this point, but it's not critical.

The focal point of this module is mitosis. Examine the four stages of mitosis and identify what is happening in each of the stages is the highlight of this module. Mitosis is only one stage in the cell cycle and should be learned that way. This is the cell division part of the module. The other three stages of the cell cycle focus on cell growth and what the cell has to do to prepare for mitosis all over again.

Once the cell cycle has been mastered, you can examine the finer points of cell division. The process of apoptosis is intriguing. All cells go through a certain number of cell divisions and then self-destruct. Examining this has to lead you to how cells go malignant. And the whole topic of cancer opens up for you to investigate as an extension.

There's also the control of cell division. Why do some cells reproduce more quickly than others? What is the driving mechanism behind this control? Imagine if all the cells of our body divided at the same time. Not only that would feel real funky, it just doesn't happen! And for good reason. Examining why is a great extension of this module.

Module 6 Goals

The following is a list of goals that should be met by the end of this module. I've made note of those that are critical, important, or nice-to-know. The critical ones can't be missed. The important ones can be missed but will only add depth if you want further study. The nice-to-know ones are exactly that. If you have the time, or your really want some depth, cover those as well.

Name the main events of the cell cycle. (**Critical**)
Describe what happens during the four phases of mitosis. (**Critical**)
Describe how the cell cycle is regulated. (**Critical**)
Explain how cancer cells are different from other cells. (Important)
Explain the problems that growth causes for cell. (Important)
Describe how cell division solves the problems of cell growth. (Important)
Identify a factor that can stop cells from growing. (Important)

Module 6 Key Words

Here's a list of key words for this module:

anaphase
apoptosis
asexual reproduction
benign
binary fission
cancer
carcinogen
cell cycle
cell differentiation
centromere
chromatid
chromatin
chromosome
cytokinesis
growth factor
histone
malignant
metaphase
metastasize
mitosis
prophase
stem cell
telomere
telophase

Module 6 Activity

CANCER STATISTICS IN THE U.S.

There are few real "labs" you can do with this module so I like to focus on a practical exercise in researching cancer statistics. It gives you a chance to practice looking up data and organizing it so it makes sense.

In this activity you are to design a chart that illustrates common characteristics of some of the most common cancers that affect Americans. When you are done you'll actually see how large the gap is between the number of people who die from the leading cause of cancer death and the one ranked #10.

CANCERS
- Bladder Cancer
- Breast Cancer
- Colon/Rectal Cancer
- Leukemia
- Lung Cancer
- Melanoma
- Non-Hodgkin's Lymphoma
- Pancreatic Cancer
- Prostate Cancer
- Uterine corpus/uterine cervix/ovary

CHART REQUIREMENTS
- List the cancers in the order of most deaths per year to least deaths per year.
- Orient in landscape mode (sideways).
- Can be hand drawn but easier to do in a spread sheet program like Excel or Numbers.
- Name the columns as follows:
  > Causes
  > Diagnostic Testing/Screening
  > Treatment

Prognosis

Deaths Per Year (U.S. Only)

- The row names are the names of the cancers.

RESOURCES

The following two web sites are a good place to start. They may not contain all the necessary information, but they are a great start.

National Cancer Institute - http://nci.nih.gov

American Cancer Society – http://www.cancer.org

Unit 2 - How Life Replicates

This unit consists of four modules and focuses on genetics, the DNA molecule and protein synthesis.

Time Frame: 6 – 10 weeks

Module 7 DNA & RNA: 2 – 3 weeks
Module 8 Introduction to Genetics: 2 – 3 weeks
Module 9 The Human Genome: 1 – 2 weeks
Module 10 Genetic Engineering: 1 – 2 weeks

Module 7: DNA & RNA

Module 7 Overview

Unit 1 covered the basic structure and function of living things. Unit 2 is going to examine the question of how life replicates. And to start, you have to have a working knowledge of the molecule of life, DNA.

In this module you need to focus on the structure of the DNA molecule. Considering the importance of this nucleic acid, the DNA structure is not all that complex. Its complexity, and the diversity of life, is in the way the individual nucleotide bases are sequenced and the length of those sequences.

What is also interesting is the fact that DNA's function is also simple to define. What is referred to as the genetic code is nothing more than the code for the production of proteins and enzymes. Nothing else. Yes, it's responsible for all our inherited traits, like hair color for example, but those traits are expressed by the presence of proteins.

To appreciate how far scientists have come in understanding the workings of our genes, and therefore DNA, it's appropriate to go back in time and examine the key contributors to the science that got us to the point where we are today. And the best place to start is with the work of Johann Friedrich Miescher. Miescher discovered the presence of an acid in the nucleus of the cell but he had no idea what is was. This discovery was done before Gregor Mendel published the Laws of Genetics and Charles Darwin published his famous book. It wasn't until the early 20th century that interest in genetics took hold, and took off, and the quest for identifying the material that controlled our inherited traits began.

So for the first half of the 20th century the race was on not only to identify DNA, through the work of Phoebus Levine, but to connect it to heredity, thanks to Oswald Avery, and identifying the molecular structure through the work of Watson and Crick.

As a really interesting extension of this, you can look in depth into the controversy surrounding the work that Watson and Crick did. There were two other researchers working on the structure of DNA, Maurice Wilkins and Roslyn Franklin. When Watson and Crick received their Nobel prize, Maurice Wilkins was also a recipient. But not Roslyn Franklin, a fact that still rankles many inside and outside science.

By 1953 the DNA molecule was successfully modeled for the first time and it wasn't the end of something. Rather, it was the beginning of another 50 years of amazing discoveries about the cell, genetics, DNA, and the human genome. We know more now about how things work than ever before. The challenge in front of us is, what do we do with this knowledge.

Since DNA is responsible for the code of proteins, you have to understand how proteins are made. For that you need to understand RNA, its structure, and the function of the different forms of RNA. You need to know where RNA is found and how it assembles proteins from the component amino acids. To do this you need to examine the processes of translation and transformation, and how RNA takes the code for the twenty-one different amino acids and makes whole protein molecules.

Module 7 Goals

The following is a list of goals that should be met by the end of
this module. I've made note of those that are critical,
important, or nice-to-know. The critical ones can't be missed.
The important ones can be missed but will only add depth if
you want further study. The nice-to-know ones are exactly that.
If you have the time, or your really want some depth, cover
those as well.

Summarize the relationship between genes and DNA.
(**Critical**)
Describe the overall structure of the DNA molecule. (**Critical**)
Summarize the events of DNA replication. (**Critical**)
Tell how RNA differs from DNA. (**Critical**)
Name the three main types of RNA. (**Critical**)
Describe transcription and the editing of RNA. (**Critical**)
Summarize translation. (**Critical**)
Describe a typical gene. (**Critical**)
Relate the DNA molecule to chromosome structure. (Important)
Identify the genetic code. (Important)
Explain the relationship between genes and proteins.
(Important)
Contrast gene mutations and chromosomal mutations.
(Important)
Describe how LOC genes are turned on and off. (Important)
Explain how most eukaryotic genes are controlled. (Important)
Relate gene regulation to development. (Important)

Module 7 Key Words

Here's a list of key words for this module:

anticodon
bacteriophage
base pairing
chromatin
codon
differentiation
DNA polymerase
exon
frameshift mutation
gene
histone
hox genes
intron
messenger RNA (mRNA)
mutation
nucleotide
operator
operon
point mutation
polyploidy
promoter
replication
ribosomal RNA (rRNA)
RNA polymerase
transcription
transfer RNA (tRNA)
transformation
translation

Module 7 Activity

EXTRACTING DNA FROM STRAWBERRIES

BACKGROUND

The wild strawberry has only two sets of chromosomes (diploid), but the grocery store variety has eight sets of chromosomes (octoploidy) and will supply a large amount of DNA. Strawberries are also soft and easy to smash, which makes them easy to work with.

Ripe strawberries produce enzymes which help in breaking down the cell walls making it easier to extract the DNA.

You will be using a solution of water, detergent and salt. This solution is called the lysis buffer. The detergent helps dissolve the lipid bilayer of the cell membranes. The salt helps remove the proteins from the DNA and keep the proteins in the extract layer so they aren't mixed in with the DNA extract.

DNA is not soluble in alcohol. When molecules are soluble, they are dispersed in the solution and are not visible. When molecules are insoluble, they clump together and become visible. The colder the alcohol, the less soluble the DNA will be in it.

After completing this lab you will have a sample of pure strawberry DNA and you will never look at strawberries in the same way ever again!

OBJECTIVES

The purpose of this lab is allow you to become familiar with a procedure for extracting DNA, collect DNA samples and observe the physical characteristics of DNA.

MATERIALS
- zip lock sandwich bag

- 1 large, ripe strawberry
- coffee filter
- large glass container (large baby food jar)
- small funnel
- lysis buffer (see instructions below)
- ethyl or isopropyl alcohol
- ice water bath
- test tube
- plastic stirrer
- scissors

PROCEDURE
1. Obtain 1 fresh strawberry. Remove any of the leaves if they are attached.
2. Place the strawberry in the zip lock bag. Squeeze all the air out and seal the bag tight.
3. Gently mash the strawberry in the bag with your fingers for at least 5 minutes. Do not break the bag.
4. At the end of 5 minutes, add 10-mL of the lysis buffer to the bag. Squeeze the air out again and seal the bag tight.
5. Mash the bag with the lysis buffer for 2 minutes. CAUTION: Mash the mixture carefully; the fewer bubbles created the better the results.
6. Place the funnel in the flask and place the coffee filter in the funnel.
7. Cut one of the bottom corners of the baggie off and slowly squeeze the strawberry liquid and the pulp into the funnel.
8. Allow the liquid to flow into the flask. Do not force the liquid.
9. Pour the filtrate into a clean test tube making sure to leave room for the alcohol.
10. Slowly pour ice cold alcohol down the side of the test tube so it forms a separate layer on top of the strawberry liquid. Do not pour the alcohol directly into the filtrate.

11. Wait about a minute and observe the formation of a thin cloudy layer in between the two liquids. This is DNA.
12. Take the plastic stirrer and gently dip the tip into the DNA fibers and slowly pull the fibers up.

LYSIS BUFFER

Lysis buffer is simple to make and is necessary for the DNA to be extracted. You need:

- 90-mL distilled water (most food stores sell distilled water in the water aisle)
- 10-mL liquid dish soap (choose your favorite brand)
- 1/4 teaspoon salt

Pour the liquid soap into the distilled water, add the salt and stir until the salt has dissolved. Your lysis buffer is ready.

ANALYSIS

A. Define: diploid, octoploidy & filtrate
B. Describe the function of DNA in living things.
C. In what part of the cell can the DNA molecule be found?
D. What is the purpose of the salt solution in this lab?
E. Why was the detergent added to the lysis buffer?
F. Why does the DNA molecule rise to the top of the test tube after the addition of the alcohol?

Module 8: Introduction to Genetics

Module 8 Overview

This module is about Mendelian genetics with a modern twist. Very little is known about the life of Gregor Mendel, the Austrian monk who discovered the principles of modern genetics. Had it not been for the fact that he published his work in an obscure journal, we never would have known he existed. After his death all his personal papers were destroyed.

As you work through how he developed the three laws of genetics, you will notice how exacting his work was and how meticulous he was about keeping statistics. It's from these records that we understand dominance, segregation of alleles, and independent assortment.

What helps in understanding Mendelian genetics is the Punnett square. This method of predicting possible outcomes of a cross is usually what people associate with when you study genetics. And predicting possible outcomes of crosses is what makes genetics interesting to many.

To understand how traits are passed on to the next generation you need to understand meiosis. Like mitosis, meiosis is cell division, but of a special kind. In mitosis, the new cells each have a normal number of chromosomes. In order for an organism to receive half its genes from each parent, mitosis will not work. With mitosis you'd end up with twice as many chromosomes, not half. Meiosis solves that problem by producing cells that have half the normal number of chromosomes. And the cells that meiosis produces are called gametes. Sex cells. As an extension, investigate the evolution of sexual reproduction. Why did organisms need to develop a method of sexual reproduction in order to insure their survival?

Once you understand meiosis you can then go on to examine crossing-over and gene mapping. You should also understand the structure of a chromosome and the terminology that goes along with those structures. Finally, before you move on to the next module, you need to understand what chromosomal abnormalities are, the types of abnormalities there are, and what causes those abnormalities.

Module 8 Goals

The following is a list of goals that should be met by the end of this module. I've made note of those that are critical, important, or nice-to-know. The critical ones can't be missed. The important ones can be missed but will only add depth if you want further study. The nice-to-know ones are exactly that. If you have the time, or your really want some depth, cover those as well.

Summarize Mendel's conclusion about inheritance. (**Critical**)
Explain the principle of dominance. (**Critical**)
Describe what happens during segregation. (**Critical**)
Describe how geneticists use Punnett squares. (**Critical**)
Explain the principle of independent assortment. (**Critical**)
Summarize the events of meiosis. (**Critical**)
Describe how Gregor Mendel studied inheritance in peas. (Important)
Explain how geneticists use the principles of probability. (Important)
Describe other inheritance patterns. (Important)
Explain how Mendel's principles apply to organisms. (Important)
Contrast the chromosome number of body cells and gametes. (Important)
Contrast meiosis and mitosis. (Important)
Identify the structures that actually assort independently. (Important)
Explain how gene maps are produced. (Nice-to-Know)

Module 8 Key Words

Here's a list of key words for this module:

allele
codominance
crossing-over
diploid
fertilization
gamete
gene
gene map
genetics
genotype
haploid
heterozygous
homologous
homozygous
hybrid
incomplete dominance
independent assortment
meiosis
multiple alleles
phenotype
polygenic trait
probability
Punnett square
segregation
tetrad
trait
true-breeding

Module 8 Activity

WORKING WITH PUNNETT SQUARES

In this exercise you will use your knowledge of genetics and conduct the same kind of experiments that Gregor Mendel did over 150 years ago. Only you're going to use the Punnett Square and no plants.

TERMINOLOGY REVIEW
Dominant Allele – when an organism always exhibits the same trait.

Recessive Allele – when an organism exhibits a trait when the dominant one is not present.

Pure Trait – a trait that has only one allele (homozygous).
Hybrid Trait – a trait that has two alleles (heterozygous).

Phenotype – what the traits actually look like.
Genotype – what genes make up the trait.

DIRECTIONS
Using the Punnett Square, determine the results for the matches listed below. For each match you are to:
- determine the phenotype of the offspring (what they look like and how many of each)
- determine the genotype of the offspring (what the alleles are and how many of each)

TRAIT	DOMINANT ALLELE	RECESSIVE ALLELE
Seed Shape	Round (R)	Wrinkled (r)
Seed Color	Yellow (Y)	Green (y)
Seed Coat Color	Gray (G)	White (g)
Pod Shape	Smooth (S)	Pinched (s)
Pod Color	Green (N)	Yellow (n)
Flower Position	Side (D)	End (d)
Height	Tall (T)	Short (t)
Flower Color	Purple (P)	White (p)

THE MATCHES

1. Cross a pure round seed shape, hybrid yellow seed color (*female traits*) with a pure round seed shape, pure green seed color (*male traits*).
2. Cross a hybrid tall, hybrid green pod color plant (*female traits*) with a hybrid tall, pure green pod color plant (*male traits*).
3. Cross a hybrid smooth pod shape, pure gray seed coat color (*female traits*) with a pure smooth pod shape and white seed coat color (*male traits*).
4. Cross a pure tall, pure round seed shape (*female traits*) with a hybrid tall, wrinkled seeds shape (*male traits*).
5. Cross a hybrid tall, hybrid gray seed coat color, pure side flower plant (*female traits*) with a pure short, pure gray seed coat color, pure end flower plant (*male traits*).

Module 9: The Human Genome

Module 9 Overview

This module builds on the previous two and focuses specifically on humans. A tremendous amount of work has been done on the human genome in the past 20 years as a result of the Human Genome Project and we know more now than we ever did before about where our genes are located on the chromosomes and a lot more about genetic disorders.

One of the first things you need to do is to understand what the human karyotype is and how we map out the 23 pairs of chromosomes. You then need to know how gender is determined in meiosis. Then you can look at the types of mutations that can occur and the kind of damage those mutations can cause.

The fun part of this module is studying pedigrees and how traits are passed down through the generations. One of the more famous genetic studies done was an investigation of the blue people of Kentucky. Studying hemophilia in the British Royal Family is interesting, but these people in Kentucky were really blue! This makes for a really cool activity in studying pedigrees.

Also in this module you need to look at some of the more common genetic disorders that humans suffer from. There are over 2000 human genetic disorders. We are aware of only about a dozen or so because of their frequency in the population. For people who have a family history of any one of the diseases, understanding the nature of the disorder is important.

There's a tremendous amount of information on the internet related to this module. You can spend as much time as you feel is appropriate here since it has such an impact on our daily lives.

Module 9 Goals

The following is a list of goals that should be met by the end of this module. I've made note of those that are critical, important, or nice-to-know. The critical ones can't be missed. The important ones can be missed but will only add depth if you want further study. The nice-to-know ones are exactly that. If you have the time, or your really want some depth, cover those as well.

Explain how small changes in DNA cause genetic disorders. (**Critical**)
Identify characteristics of human chromosomes. (**Critical**)
Summarize non-disjunction and the problems it causes. (**Critical**)
Summarize methods of human DNA analysis. (**Critical**)
Identify the types of human chromosomes in a karyotype. (Important)
Explain how sex is determined. (Important)
Explain how pedigrees are used to study human traits. (Important)
Describe examples of the inheritance of human traits. (Important)
Describe some sex-linked disorders and explain why they are more common in males than in females. (Important)
Describe how researchers are attempting to cure genetic disorders. (Important)
State the goal of the Human Genome Project. (Nice-to-Know)
Explain the process of X-chromosome inactivation. (Nice-to-Know)

Module 9 Key Words

Here's a list of key words for this module:

autosome
DNA fingerprinting
karyotype
nondisjunction
pedigree
sex chromosome
sex-linked gene

Module 9 Activity

YOUR GENES YOUR HEALTH

Use the web site "Your Genes Your Health", www.ygyh.org, to answer the following questions about the listed genetic disorders.

Down Syndrome
- How is the mosaic form of Down Syndrome (DS) different from the nondisjunction form?
- Which type of DS can be inherited?
- What are the chances that a man with DS will have a baby? A woman? Why are there differences?

Sickle Cell Anemia
- What happens when hemoglobin polymers form?
- How does hemoglobin electrophoresis diagnose people with sickle cell?
- How does gel electrophoresis show the difference between a baby with sickle cell and a healthy baby?

Hemophilia
- What is the difference in clotting between people with severe Hemophilia A and mild Hemophilia A?
- What makes Hemophilia-A different from Hemophilia-B?
- How does X-inactivation lead to hemophilia in girls?

Cystic Fibrosis
- What is the reason for the mucus in the pancreas of a CF patient?
- In cases where the sweat test for CF can't be done, how does the DNA test confirm CF?
- Draw the Punnett square of a man with CF and a carrier female and explain the odds of their child having the disease?

Module 10: Genetic Engineering

Module 10 Overview

This last module in Unit 2 will round out where we have been with genetics, where we are in genetics today, and what the future holds. Genetic engineering is considered by many biologists to be the future of biology. But the social, political, economic, and religious implications are huge.

Start off by examining selective breeding and its history. Animal husbandry has been going on for hundreds of years. Farmers and ranchers are always looking for a better yielding crop resistant to disease as the rancher is looking for getting as much meat from the herd as possible. Selective breeding has helped advance agriculture to the science we see today. For example, the original potato looks nothing like what we eat today. The original is small and purple and is still cultivated in the Andes mountains of South America. The chips have come a long way!

As you examine how selective breeding has lead to genetic engineering, you must look at DNA analysis and what that means, not so much as a crime fighting tool, but as a tool to literally build a better potato. Now you get into the controversy of genetically modified foods. And a good side trip to this is looking at gene patents. Who owns your genes? Do you? Think again.

Farmers in the U.S. have already faced this with the seeds they buy and plant. It's well known that practically all the soy bean seeds sold in the U.S. are sold by Monsanto. They have produced a genetically modified seed and they hold the patent on that seed. You cannot plant soy beans from the seeds produced by the Monsanto seed. And Monsanto aggressively seeks out those farmers who they believe are infringing on their patent and takes them to court. They usually win. Corn is the other big crop that is also genetically modified.

From plants you can make the jump to animals and discover a similar kind of controversy. Consider all the cows, pigs, and chickens we eat and ask yourself where they all come from and how is it they all taste alike. The final connection you need to make is the human connection.

Genetic engineering is the key to treating and curing many of the devastating hereditary diseases humans suffer from. From a humanitarian viewpoint this can be a good thing. But with many good things of this nature, there is also a darker side. Back in the 1930s eugenics was a popular theory. Breeding out undesirable human traits and replacing them with desirable ones. Becoming part of a political agenda is a nightmare, and we experienced that in our struggle to overcome fascism in the 1940s.

Finally, consider the impact of a simple DNA fingerprint. Members of the U.S. Military are required to have their DNA fingerprint on file. This is for identification purposes in case a body cannot be identified through normal means. What if this information is leaked or sold to insurance companies. If you have a family history of a disease could they deny you coverage? Or even employment if employers have access to this information? This is scary stuff to consider. And the reality of these scenarios can happen overnight.

Module 10 Goals

The following is a list of goals that should be met by the end of this module. I've made note of those that are critical, important, or nice-to-know. The critical ones can't be missed. The important ones can be missed but will only add depth if you want further study. The nice-to-know ones are exactly that. If you have the time, or your really want some depth, cover those as well.

Explain the purpose of selective breeding. (**Critical**)
Tell why breeders try to induce mutations. (**Critical**)
Summarize what happens during transformation. (**Critical**)
Explain how scientists manipulate DNA. (Important)
Explain how you can tell if a transformation experiment has been successful. (Important)
Describe the usefulness of some transgenic organisms to humans. (Important)
Describe two techniques used in selective breeding. (Important)
Summarize the main steps in cloning. (Nice-to-know)

Module 10 Key Words

Here's a list of key words for this module:

clone
gel electrophoresis
genetic engineering
genetic marker
hybridization
inbreeding
plasmid
polymerase chain reaction (PCR)
recombinant DNA
restriction enzyme
selective breeding
transgenic

Module 10 Activity

TRANSGENEIC FLY VIRTUAL LAB

Genetic engineering is one of the "hot" topics in modern biology from genetically modified foods to curing genetic disorders through gene replacement therapy.

The best way to demonstrate this is through a virtual lab exercise. The Howard Hughes Medical Institute provides a number of excellent virtual labs that are not only enjoyable, but educational. The Transgeneic Fly Virtual lab is one of the labs found at the HHMI web site and it does an excellent job at allowing you to experience what one form of genetic engineering is all about.

Here's the web address:
http://www.hhmi.org/biointeractive/vlabs/transgenic_fly/index.html.

Have fun making flies!

Unit 3 - How Life Adapts

This unit consists of four modules and focuses on evolution, the history of life, and classification.

Time Frame: 6 – 10 weeks

Module 11 History of Life: 2 weeks
Module 12 Darwinian Evolution: 3 weeks
Module 13 Evolution of Populations: 2 weeks
Module 14 Classification: 1 week

Module 11: History of Life

Module 11 Overview

This first module of Unit 3 is the Earth's time capsule. There's an old saying that goes, "How do you know where you are going if you don't know where you've been?" In this case, how can we understand today's living world if we don't know where it all came from?

The History of Life traces the origin of life on the planet and should trace the origin of the planet itself. As we continue to explore the galaxy and the Universe, we continually discover more and more evidence about the origins of pretty much everything.

This module is unique in that it's a good mix of biology and geology, with a whole bunch of paleontology thrown in to make it interesting. What's key in this module is understanding the accepted scientific theories of how the solar system was formed over 5 billion years ago, how the Earth was formed over 4.5 billion years ago, and how life first emerged about 3.5 billion years ago.

In order to make this module more understandable, and easier to grasp, it's best to deal with these big chunks of time as they are set up in the geologic time line. By examining eras and periods from Earth's geologic history, you can overlay the history of life since the evolution of life is tied to geologic upheavals, mass extinctions, and those pesky drifting continents. When you add ice ages and climate changes, then you'll understand why life on this planet is not only diverse, but sometimes confined to specific geologic areas.

Along the way you'll discover how humans and dinosaurs couldn't have possibly existed together (much to the disappointment of all you Jurassic Park fans) and why it's believed that the dinosaurs may have already been on the road to extinction well before we got hit with that asteroid 65 million years ago.

And of course there are those mass extinctions themselves. We know there have been five mass extinctions which have resulted in life changing on Earth forever. The last one, 65 million years ago, allowed mammals to become dominant, which eventually led to the rise of humans. Wonder what the next mass extinction will bring. It's not a matter of if it will happen, but when.

Module 11 Goals

The following is a list of goals that should be met by the end of this module. I've made note of those that are critical, important, or nice-to-know. The critical ones can't be missed. The important ones can be missed but will only add depth if you want further study. The nice-to-know ones are exactly that. If you have the time, or your really want some depth, cover those as well.

Describe the fossil record. (**Critical**)
State the hypotheses that have been proposed for how life first arose on Earth. (**Critical**)
Identify some of the main evolutionary steps in the early evolution of life. (**Critical**)
Identify the divisions of the geologic time scale. (**Critical**)
Describe how conditions on early Earth were different from conditions today. (Important)
State the information that relative dating and radioactive dating provide about fossils. (Important)
Identify important patterns of macroevolution. (Important)
Explain what Miller & Urey's experiments showed. (Nice-to-Know)

Module 11 Key Words

Here's a list of key words for this module:

adaptive radiation
coevolution
convergent evolution
endosymbiotic theory
era
extinct
fossil record
geologic time scale
half-life
index fossil
macroevolution
mass extinction
microfossil
paleontologist
period
proteinoid microsphere
punctuated equilibrium
radioactive dating
relative dating

Module 11 Activity

HISTORY OF LIFE TIME LINE

BACKGROUND
About 4.5 billion years ago, Earth was a ball of hot, molten rock. As the surface of the planet cooled, a rocky crust formed and water vapor in the atmosphere condensed to form rain. By 3.9 billion years ago, oceans covered most of the Earth's surface. Rocks formed in these oceans contain fossils of bacterial cells that lived about 3.5 billion years ago. The fossil record shows a progression of life-forms and contains evidence of many changes in Earth's surface and atmosphere.

OBJECTIVES
The purpose of this lab is to organize the appearance of life on Earth in a timeline. This timeline can be used to study how living things have changed over time.

MATERIALS
- adding-machine tape 1 1/2 – 2 inches wide
- metric measuring tape
- 12" metric ruler
- pencil
- colored markers/pencils

RESOURCE
You will need a geologic time reference. Here's a good start:
http://en.wikipedia.org/wiki/Timeline_of_evolution

PROCEDURE
1. Obtain a 2.5-meter length of cash register tape.
2. Carefully stretch out the tape on a table, or the floor, for its entire length.
3. Starting at the left-hand end of the tape, make a pencil mark every 10-cm.
4. After completing step #3, label the left-hand end of the

tape "5 Billion Years Ago", and the other end "Today."

5. Write "10-cm = 200 million Years" near the beginning of the tape.
6. Label the points on the tape that represent 4 billion, 3 billion, 2 billion and 1 billions years.
7. Label the point on your timeline that represents the formation of Earth.
8. Refer to the above web resource and label the points on the timeline that represent the following events:
   > First bacteria appearing
   > Oxygen entering the atmosphere
   > Eukaryotes appear
   > Multi-cellular organisms appear
   > The five mass extinctions
   > The first dinosaurs appear
   > The first plants appear
   > The first humans appear
9. Label the 11 periods of geologic time scale. Write the name of the period and the range of years on the tape.

ANALYSIS

Using the scale from step #1 where 10-cm = 200 million years, you can use the analogy of a 24-hour clock to place major events. Starting with 4.8 billion years as midnight, answer the following questions using hours as your answers.

A. How long has life existed on earth?
B. For what part of the day did only unicellular life-forms exist?
C. At what time of the day did the first plants appear on Earth?
D. At what time of the day did mammals appear on Earth.
E. Identify the major developments in life-forms that have occurred over the last 3.5 billion years.
F. How do mass extinctions appear to be related to the appearance of new major groups of organisms?

G. Photosynthetic bacteria are thought to be responsible for adding oxygen to the Earth's atmosphere. Justify this conclusion with evidence from the time line.

Module 12: Darwinian Evolution

Module 12 Overview

The single most important goal of every species is to prevent their own extinction. Populations of organisms do not collectively decide that they have existed long enough and therefore have to become extinct. How species prevent extinction and the explanation of that process is the topic of this module. Biological evolution is the mechanism by which species can prevent their own extinction. And the man who gave us this explanation, in part, was Charles Darwin.

To truly understand biological evolution according to Darwin's theories, it's really important to examine the works of those scientists who preceded Darwin and came up with their own theories of evolution. Scientists like George Cuvier, Jean-Baptiste Lamark, and George Leclerc each developed a theory of evolution that pre-dated Darwin's work by decades. When you add the geologists James Hutton and Charles Lyell to the mix, you get the coming together of all these separate ideas into a focused theory that made sense. And this is what Darwin accomplished as he made his famous around the world tour in the H.M.S. Beagle. The only thing missing from Darwin's work was the genetic explanation of natural selection provided through the work of Gregor Mendel. Biological evolution had to wait until the 20th century for genetics and evolution to begin melding together.

There are two key principles in evolution; descent with modification and natural selection. Both of these principles are explained by Darwin in his book, *On the Origin of Species by Natural Selection*. It's also important to understand what Darwin did not say in his writings about evolution. For example, he did not say that man descended from apes, monkeys, or chimps. He did not say that evolution was an end point or that organisms evolved from simpler life forms. And he did not coin the term "survival of the fittest." He gets the blame for all this, and more, but he did not say any of it.

Biological evolution is the thread that binds all of biology together. When you step back and look at the characteristics of life, they are all designed to support the successful reproduction of the species. Without successful reproduction in sufficient numbers, species will become extinct, regardless of any genetic modifications they have made to survive environmental changes.

This module is an important one in the understanding of modern biology. Without it, all the other modules just become free-standing topics with nothing holding the entire thing together.

Module 12 Goals

The following is a list of goals that should be met by the end of
this module. I've made note of those that are critical,
important, or nice-to-know. The critical ones can't be missed.
The important ones can be missed but will only add depth if
you want further study. The nice-to-know ones are exactly that.
If you have the time, or your really want some depth, cover
those as well.

Describe how natural variation is used in artificial selection.
(**Critical**)
Explain how natural selection is related to species' fitness.
(**Critical**)
Identify evidence Darwin used to present his case for evolution.
(**Critical**)
State Darwin's theory of evolution by natural selection.
(**Critical**)
Describe the pattern Darwin observed among organisms of the
Galapagos Islands. (Important)
State how Hutton and Lyell described geological change.
(Important)
Identify how Lamark thought species evolve. (Important)
Describe Malthus' theory of population growth. (Important)
List the events leading to Darwin's publication of *On the
Origin of Species.* (Nice-to-Know)

Module 12 Key Words

Here's a list of key words for this module:

adaptation
artificial selection
common descent
descent with modification
evolution
fitness
fossil
homologous structures
natural selection
struggle for existence
survival of the fittest
theory
vestigial organ

Module 12 Activity

FOUNDERS OF EVOLUTION

In this activity you are going to draw a time line of the lives of the 11 most influential people in evolution. This time line is going to cover a 200-year period in the history of evolution. When you are done you will be able to see how the lives of many of these people over-lapped.

MATERIALS
- 140-centimeter length of cash register tape, 1 1/2-2 inches wide
- meter stick
- 12-inch metric ruler
- pencil
- colored pencils/markers

PROCEDURE
1. Obtain the 140-cm length of cash register tape, the meter stick and a ruler.
2. Using the meter stick, mark off 5-cm sections.
3. With a pencil, draw a light vertical line separating each 5-cm section.
4. You should have 22 sections on your tape, each 5-cm wide.
5. Each section represents a decade, starting with 1700 and ending with 1920.
6. At the top of the tape, on each line separating the sections write the decade, starting with 1710. The last section should have 1910.
7. You are going to draw a line that represents the lifespan of each person. You must start with the earliest possible date and work from there. To accommodate you, the list below has been assembled in the order in which you are to draw the lines.
8. With the colored pencils draw a line for each scientist.

Start the line with their year of birth and stop with their year of death.

9. HINT: You have enough space for the first 9 names. The last two (Darwin & Wallace) can be drawn at the same level as the others since both of their birth years are late enough not to interfere with any other line. (Refer to the sample)

10. Leclerc's line should be as close to the top of the tape as possible.

11. Space out the next 8 lines as evenly as possible. Lines cannot overlap!

12. You are to write the name of the scientist at the beginning or at the end of the line. Be sure to include their birth and death years.

FOUNDERS LIST

George Louis Leclerc 1707-1788	Adam Sedgwick 1785-1873
James Hutton 1726-1797	Charles Lyell 1797-1875
Erasmus Darwin 1731-1802	John Henslow 1796-1861
Jean-Baptiste Lamarck 1744-1829	Charles Darwin 1809-1882
Thomas Malthus 1766-1834	Alfred Wallace 1823-1913
George Cuvier 1769-1832	

Module 13: Evolution of Populations

Module 13 Overview

When Darwin published his book in 1859, Gregor Mendel was still seven years away from publishing his now famous principles of genetics. It's ironic that the two men who changed biology and biological thought forever didn't know of each other's existence and who unknowingly needed each other's work to help explain their ideas. This module brings together Darwin's and Mendel's work and examines the mechanisms behind the prevention of extinction. In other words, how species are formed.

To start with, you need to examine those conditions that must be present in order for new species to form. Usually these conditions are environmental in nature. From here you need to go into the mechanisms of natural selection and study how new traits are selected and how they become an advantage to the population.

As this progression continues, the selected traits may be single-gene or polygenic. These should have been studied already and now you get to see how they work in a population. Of course no study of populations is complete without knowing what the gene pool is. We use this term in our everyday language, but now you will see how it applies in biology. What does it really mean and how does it work?

Lastly, no study of the gene pool is complete without knowing what genetic drift is and how it affects the gene pool. The Hardy-Weinberg principle is an important aspect of this explanation.

Darwin spent a great deal of time discussing how species became species. He knew it was tied to inheritance in some way, but he didn't know what the mechanism was. He died three decades before Mendel's work was discovered and accepted. It was decades more before the mechanisms of evolution were explained by geneticists. Now when we look back in time and see the rise and fall of the dominance of life forms, we have the means to understand not only the whys, but the hows as well.

Module 13 Goals

The following is a list of goals that should be met by the end of this module. I've made note of those that are critical, important, or nice-to-know. The critical ones can't be missed. The important ones can be missed but will only add depth if you want further study. The nice-to-know ones are exactly that. If you have the time, or your really want some depth, cover those as well.

Explain what a gene pool is. (**Critical**)
Identify the condition necessary for a new species to evolve. (**Critical**)
Explain how natural selection affects single-gene and polygenic traits. (**Critical**)
Describe genetic drift. (**Critical**)
List the five conditions needed to maintain genetic equilibrium. (Important)
Describe the process of speciation in the Galápagos finches. (Important)
Identify the main sources of inheritable variation in a population. (Important)
State what determines how a phenotype is expressed. (Important)

Module 13 Key Words

Here's a list of key words for this module:

behavioral isolation
directional selection
disruptive selection
founder effect
gene pool
genetic drift
genetic equilibrium
geographic isolation
Hardy-Weinberg principle
polygenic trait
relative frequency
reproductive isolation
single-gene trait
speciation
stabilizing selection
temporal isolation

Module 13 Activity

HOW DOES GENETIC DRIFT OCCUR?

BACKGROUND
Use a deck of cards to represent a population of island birds.
The four suits represent different alleles for tail shape. The
allele frequencies in the original population are 25% spade,
25% heart, 25% club, and 25% diamond tail shapes.

MATERIALS
- Deck of cards

PROCEDURE
1. Shuffle the cards and hold the deck face down. Turn
   over 40 cards to represent the alleles of 20 offspring
   produced by random matings in the initial population.
2. Separate the 40 cards by suit. Find the allele
   frequencies for the offspring by calculating the
   percentage of each suit.
3. Suppose a storm blows a few birds to another island.
   They are isolated on this island and start a new
   population. Reshuffle the deck and draw 10 cards to
   represent the alleles of five offspring produced in the
   smaller population.
4. Repeat step 2 to calculate the resulting allele
   frequencies.

ANALYZE AND CONCLUDE
A. Compare the original allele frequencies to those
   calculated in steps 2 and 4. How did they change?
B. Did step 1 or 3 demonstrate genetic drift?
C. Does this activity demonstrate evolution? Why or why
   not? Does it demonstrate natural selection? Explain.

Module 14: Classification

Module 14 Overview

Scientists, especially biologists, love to classify things. Everything must be in a category or there is scientific chaos! With the wide diversity of life present on our planet, organizing all that life into categories only makes sense. The science of classification is called taxonomy and that is the topic of this last module in Unit 3.

The study of classification has a long history in biology and started with Hippocrates. And for many centuries following Hippocrates, classification was based on structure. If similar structures were found in similar organisms, they were classified together.

The first accepted classification scheme was developed in the 18th century by the Swedish botanist Linnaeus. The foundations of that scheme are still in place now, though highly modified and detailed. Understanding the hierarchy of the Linnaean system is important to understanding the modern system of classification. You need to know what the categories and sub-categories are.

Up until the mid-twentieth century, all life was classified into two kingdoms. But there were problems with this system. There were organisms that didn't quite fit into either the plant kingdom or the animal kingdom. Studying how the two kingdom system grew into the six kingdom system we have today is important to understanding taxonomy.

A significant improvement in classification came as a result of DNA analysis. Now we have the ability to look at an organism's genes and classify, and in some cases re-classify, organisms according to their similarities in genes, and not solely structure. That also allows us to study evolutionary relationships as a basis for classification.

The modern classification scheme considers evolutionary relationships through the study of cladistics. Cladistic analysis allows us to see how modern organisms are related to their extinct relatives and is a critical part of the study of classification.

Module 14 Goals

The following is a list of goals that should be met by the end of this module. I've made note of those that are critical, important, or nice-to-know. The critical ones can't be missed. The important ones can be missed but will only add depth if you want further study. The nice-to-know ones are exactly that. If you have the time, or your really want some depth, cover those as well.

Explain how living things are organized. (**Critical**)
Describe binomial nomenclature. (**Critical**)
Explain how evolutionary relationships are important in classification. (**Critical**)
Identify the principle behind cladistic analysis. (**Critical**)
Explain how we can compare very dissimilar organisms. (Important)
Name the six kingdoms. (Important)
Describe the three-domain system of classification. (Important)
Explain Linnaeus' system of classification. (Important)

Module 14 Key Words

Here's a list of key words for this module:

Animalia
Archaea
Archaebacteria
bacteria
binomial nomenclature
cladogram
class
derived character
domain
Eubacteria
Eukarya
evolutionary classification
family
Fungi
genus
kingdom
molecular clock
order
phylogeny
phylum
Plantae
Protista
taxon
taxonomy

Module 14 Activity

NAME THAT POTATO CHIP

There are many layout styles for a dichotomous key. This is just one example of many possibilities!
The population of potato chips used:
- Lay's Classic
- Lay's BBQ
- Ruffles BBQ
- Ruffles Original
- Pringles Original
- Lay's Stax Sour Cream and Onion
- Pringles Cheddar Cheese

Options:
Add more subtle distinctions, such as thick cut chips or baked chips which would require some kind of measurement to distinguish between types.
Change the population to include all "chips" such as, corn chips, vegetable chips, etc.

Unit 4 - How Life Interacts

This unit consists of four modules and focuses on ecology and ecological issues.

Time Frame: 6 – 10 weeks

Module 15 The Biosphere: 1 week
Module 16 Ecosystems & Communities: 2 weeks
Module 17 Populations: 2 weeks
Module 18 Human Impact in the Environment: 3 weeks

Module 15: The Biosphere

Module 15 Overview

In this module on the biosphere you need to examine the organization of living things not based on individual species, like you did in classification, but the whole world. Mother Earth is divided into four spheres. The biosphere is the one that includes all living things and their non-living habitats.

If you take the traditional approach you will travel down the pyramid from biosphere through ecosystem to biomes and finally to habitats. Then, of course, you have to consider the predator-prey relationships that exist. What I like to call "who's eating whom." Then you get into producer/consumer relationships, who's the omnivore, herbivore, carnivore, and detritivore. Then you have to get into the "-isms." Commensalism, parasitism, mutualism and symbiosis (I know, it's not an "ism" but an "osis," but it still belongs).

There are food webs and food chains and biogeochemical cycles to consider also. The really important ones are the carbon cycle, water cycle, the oxygen, and nitrogen cycle.

Eventually you'll get to trophic levels. These are where the energy is located. I suggest that you start here. As the FBI is fond of "following the money trail" to solve crimes, you should follow the energy trail if you want to understand ecosystems. It's all about the flow of energy through a system that makes the ecosystems, biomes, habitats unique. It's the availability of energy that will allow species to be successful in any given habitat or biome.

You need to understand the role of photosynthesis at this level and view its importance to the biosphere differently. Once you've accomplished that, then energy pyramids and trophic levels make more sense. Now you can examine any ecosystem and understand why they are specific to the types of organisms that live there.

It's all about the flow of energy—who gets it and how it's used. And how it's used is the easy part if you understand that in order to prevent their own extinction a population needs to reproduce successfully. In order to do that they need an accessible source of, you guessed it, energy. And that accessible source of energy can be found where they live.

Module 15 Goals

The following is a list of goals that should be met by the end of
this module. I've made note of those that are critical,
important, or nice-to-know. The critical ones can't be missed.
The important ones can be missed but will only add depth if
you want further study. The nice-to-know ones are exactly that.
If you have the time, or your really want some depth, cover
those as well.

Identify the levels of organization that ecologists study.
(**Critical**)
Identify the source of energy for life processes. (**Critical**)
Trace the flow of energy through living systems. (**Critical**)
Evaluate the efficiency of energy transfer among organisms in
an ecosystem. (**Critical**)
Describe how matter cycles among the living and nonliving
parts of an ecosystem. (**Critical**)
Explain why nutrients are important in living systems.
(Important)
Describe how the availability of nutrients affects the
productivity of ecosystems. (Important)
Describe the methods used to study ecology. (Important)

Module 15 Key Words

Here's a list of key words for this module:

algal bloom
autotroph
biogeochemical cycle
biomass
biome
biosphere
carnivore
chemosynthesis
commensalism
community
consumer
decomposer
denitrification
detritivore
ecological pyramid
ecology
ecosystem
evaporation
food chain
food web
herbivore
heterotroph
limiting nutrient
mutualism
nitrogen fixation
nutrient
omnivore
parasitism
photosynthesis
population
predator-prey relationship
primary productivity
producer

Module 15 Key Words

species
symbiosis
transpiration
trophic level

Module 15 Activity

ABIOTIC FACTORS AND PLANT GROWTH

Many factors affect plant growth. Is it possible to test some in a laboratory setting? In this investigation you will choose an abiotic factor and attempt to test how (or if) it affects the growth of radish seedlings.

PROBLEM
How do abiotic factors affect plant growth?

MATERIALS
- 4 radish seedlings
- 4 cups
- ruler
- cheesecloth
- sand
- gravel
- potting soil
- household-plant liquid fertilizer
- plastic wrap in a variety of colors
- graduated cylinder

PROCEDURE
1. Choose an abiotic factor to test on the growth of radish seedlings. Possible factors include amount of sunlight, amount of water, soil type, light color available to plants, or amount of fertilizer.
2. Determine a way to vary the factor you have chosen. Be sure to include at least three different settings of your variable and to keep all other factors constant. Write out a procedure for your investigation.
3. Obtain 4 plants. Label one "Control" and the remaining three "A," "B," and "C."
4. Measure the height of your control and variable plants over a period of seven days. Use the same method to

repeat measurements each day. Be sure to keep plants watered.

5. Record all data you generate in a well-organized data table.

ANALYZE AND CONCLUDE

A. On the basis of your procedure, how are you defining plant growth?
B. What are your independent and dependent variables? What are your constants? What is your control?
C. Make a bar graph to present the data you obtained on plant growth.
D. By studying your data, what can you conclude about how (or if) your variable affects the growth of radish seedlings?
E. Is your experiment a failure if your variable did not apparently affect the growth? Explain.
F. What possible sources of error may have occurred in your experiment? Why might they have occurred?

EXTENSION

How would you design an experiment to determine whether a specific biotic factor influences plant growth?

Module 16: Ecosystems & Communities

Module 16 Overview

This module is an in-depth extension of the previous one. In this module you need to differentiate the various ecosystems found on the planet and the biomes that make up those ecosystems. For a real challenge go down to the community level!

To understand this, you need to study ecological succession and the types of ecological succession that are found. There is indeed a pattern that has been observed and all ecosystems will follow these patterns.

Climate is a big factor in how ecological succession takes place, so you need to understand what climate is and what differentiates climates like temperate, arctic, or tropical. You also need to know how climates change over a long period of time and the impact that climate change has on ecosystems. This is a critical point to understand if you're going to honestly look at human impact on ecosystems and the controversy surrounding things like "global warming" and "climate change."

Don't be misled by such a short overview of this module. There is great depth here. An examination of ecosystems, biomes and communities is a challenge. When you add to that a serious consideration of climate, you've got the core of an exciting and socially relevant topic.

Module 16 Goals

The following is a list of goals that should be met by the end of this module. I've made note of those that are critical, important, or nice-to-know. The critical ones can't be missed. The important ones can be missed but will only add depth if you want further study. The nice-to-know ones are exactly that. If you have the time, or your really want some depth, cover those as well.

Identify the causes of climate. (**Critical**)
Explain how biotic/abiotic factors influence an ecosystem. (**Critical**)
Identify the characteristics of major land biomes. (**Critical**)
Identify the factors that govern aquatic ecosystems. (**Critical**)
Identify the two types of freshwater ecosystems. (**Critical**)
Describe the characteristics of the marine zones. (**Critical**)
Identify interactions that occur within communities. (Important)
Describe how ecosystems recover from a disturbance. (Important)
Explain what microclimates are. (Important)
Explain how Earth's temperature range is maintained. (Important)
Identify Earth's three main climate zones. (Important)

Module 16 Key Words

Here's a list of key words for this module:

abiotic factor
aphotic zone
benthos
biome
biotic factor
canopy
climate
coastal ocean
competitive exclusion principle
coniferous
coral reef
deciduous
detritus
ecological succession
estuary
greenhouse effect
habitat
humus
kelp forest
mangrove swamp
microclimate
niche
permafrost
photic zone
phytoplankton
pioneer species
plankton
polar zone
predation
primary succession
resource
salt marsh
secondary succession

Module 16 Key Words

taiga
temperate zone
tolerance
tropical zone
understory
weather
wetland
zonation
zooplankton

Module 16 Activity

SAVE THE RAINFOREST

THE SITUATION
The government has decided that it needs the resources of the rainforests and is going to cut them all down. Unless the government can be convinced that it cannot, must not, do this, the rainforests will disappear forever.

You don't think this is a good thing to do and you want to stop them. You must convince the government that they are making a big mistake. You must convince the government that their decision will affect many things.
In order to stop the government, you need to show the government that the rainforests are important to the whole earth. You will be able to do this by completing the following tasks.

THE PROJECT
You work to convince the government that their decision to eliminate the rainforests is a bad one. You will accomplish this by working through the following steps.

You will design a poster that will convince the government to stop their plans to destroy the rainforests.

The poster must have the following items on it:
• a slogan that will get attention
• pictures of the endangered plants and animals in the rainforest
• a map of the location of the forest which will include a compass rose and a map key
• a typical food web or energy web of the rainforest

The poster cannot be smaller than construction-sized paper and be done on oak tag-type material.

The poster will be done in color.

THE RESOURCES
You have access to many resources. In order for your poster to be convincing, you must know all you can about rainforests, where they are located, the kinds of plants and animals that are there, and the climates that rainforests are located in.

To do this, you need to do research. The following web sites are all current and will give you as much information you need to decide if you agree or disagree with cutting down the rainforests. Remember, your poster must be able to convince whoever is looking at it of your position.

WEB SITES
Dr. Blythe's Rainforest Education web site –
http://www.rainforesteducation.com

Exploring the Environment – http://www.cotf.edu/ete/

How To Save Tropical Rainforests –
http://www.mongabay.com/1001.htm

How Rainforests Work –
http://science.howstuffworks.com/rainforest.htm

North American Association for Environmental Education –
http://www.naaee.org

Rainforest Action Network – http://www.ran.org/info_center

Rainforest Facts – http://www.rain-tree.com/facts.htm

Where Are The Rainforests? –
http://www.srl.caltech.edu/personnel/krubal/rainforest/Edit560s6/where.html

John Turano

Module 17: Populations

Module 17 Overview

Once you understand how energy flows through an ecosystem, and what biomes are, you're ready to look at how populations increase, or decrease, in any given habitat.

To do this you need to understand what a population needs in order to increase in size, both the biotic and abiotic factors. Then you need to examine those factors that are solely dependent on population size and those that are independent of population size. In other words, there are things that a population can control and there are things that it cannot.

These limiting factors will enable you to understand concepts like carrying capacity, logistic growth, and exponential growth.

Once you understand these critical concepts, use them to study human population growth. There are nearly seven billion of us on the planet. No one really knows where it will stop, or when. But as food gets scarce, and drinkable water get scarce, people are going to die not only from hunger and thirst, but they are going to put a great deal of pressure on their governments to either get food and water or protect what food and water they have. And that pressure could have far reaching political and economic impact for many countries.

Human population control is very controversial, to say the least, but if you understand the mechanics of population growth in general, then you'll be better armed to make informed, intelligent decisions.

Module 17 Goals

The following is a list of goals that should be met by the end of
this module. I've made note of those that are critical,
important, or nice-to-know. The critical ones can't be missed.
The important ones can be missed but will only add depth if
you want further study. The nice-to-know ones are exactly that.
If you have the time, or your really want some depth, cover
those as well.

Identify factors that affect population size. (**Critical**)
Differentiate between exponential and logistic growth.
(**Critical**)
Identify factors that limit population growth. (**Critical**)
Differentiate between density-dependent and density-
independent limiting factors. (**Critical**)
Describe how the size of the human population has changed
over time. (Important)
List the characteristics used in describing a population.
(Important)
Explain why population growth rates differ in countries around
the world. (Nice-to-Know)

Module 17 Key Words

Here's a list of key words for this module:

age-structure diagram
carrying capacity
demographic transition
demography
density-dependent limiting factor
density-independent limiting factor
emigration
exponential growth
immigration
limiting factor
logistic growth
population density

Module 17 Activity

POPULATION BIOLOGY

Go to the Population Reference Bureau's web site (www.prb.org) and download the document called, "2011 World Population Data Sheet". The following questions are based on data found in the reference sheet.

1. What is the current population of the world?
2. Rank the ten countries with the largest population (from largest to smallest).
3. Which country had the fewest number of girls than boys enrolled in secondary schools in 1990? In 2005?
4. The crude death rate (CDR) is the annual number of deaths per 1,000 population. Which Region has the highest CDR? Which Region has the lowest?
5. The infant mortality rate measures the number of deaths each year to infants under one year of age per 1,000 live births. Which Region has the highest infant mortality rate and what is that rate? Which Region has the lowest and what is that rate?
6. The total fertility rate (TFR) is the average number of children women would have if they maintained the current level of childbearing throughout their reproductive years. Which countries share the highest TFR and what is it? Several countries share the lowest TFR. What is the lowest TFR?
7. Which countries have the "youngest" population, that is, the highest proportion of population under age 15?
8. Which country has the "oldest" population, that is, the highest proportion of population over age 64?
9. In which country are people expected to live the longest? Which country has the lowest life expectancy?
10. A population grows because there are more births than deaths or more people are moving in than moving out. The difference between births and deaths is expressed

as a percentage called the rate of natural increase. Which region is growing the fastest through natural increase? Which region is growing at the slowest rate?

11. Which Continent is growing the fastest through natural increase? Which Continent is growing at the slowest rate?

12. Rank the ten countries with the largest projected populations for both 2025 and 2050 (from largest to the smallest).

13. Which country has the highest adult HIV/AIDS percent for 2007/2008? How many countries share the lowest rate and what is the rate?

14. Rank the Regions according to population size (from largest to the smallest).

Module 18: Human Impact in the Biosphere

Module 18 Overview

This last module will tie the previous three together and examine man's impact on the environment. And it fits hand-in-glove with the module on populations. As the human population continues to grow, the strain that the increased population has on resources, both man-made and natural, become greater and greater.

You want to examine the different types of pollution that exist, both man-made and natural. Nothing pollutes the air more than a massive volcanic eruption! When Krakatoa blew its top in 1883, it spewed so much pollution into the atmosphere that the world's climates changed for years. *That* is climate change!

You also have to examine the differences between conservationism and environmentalism. Though the line between the two can blur, one is a Nature movement and one is a political movement. And they all have a political agenda.

You need to know what a sustainable and a renewable resource is and how recycling can help protect natural resources. What does "going green" really mean? And is it having a positive impact?

By the time you get to this module, you should have a good understanding that everything is connected to everything else on the planet. We can no more separate the biosphere from the geosphere as we can separate the digestive system from the circulatory system. We're all connected in some way and since Nature abhors imbalance, what goes on in one part of the world will have an effect in another. When the oceans can no longer absorb all the carbon dioxide that is generated from plants, animals, and human activity, what happens then not only to the atmosphere, but to all living things that depend on oxygen as a source of energy?

Of all the modules, this is where you should use your knowledge of biology to examine the critical ecological issues that the world faces. And in your own way, it's my hope that you'll pay it forward by getting others to see how important understanding the biological world is.

Module 18 Goals

The following is a list of goals that should be met by the end of this module. I've made note of those that are critical, important, or nice-to-know. The critical ones can't be missed. The important ones can be missed but will only add depth if you want further study. The nice-to-know ones are exactly that. If you have the time, or your really want some depth, cover those as well.

Explain how environmental resources are classified. (**Critical**)
Describe how human activities affect land, air, and water resources. (**Critical**)
Define biodiversity and explain its value. (**Critical**)
Identify current threats to biodiversity. (**Critical**)
Describe two types of global change that are of concern to biologists. (**Critical**)
Describe the goal of conservation biology. (Important)
Identify the characteristics of sustainable development. (Important)
Describe human activities that can affect the biosphere. (Important)

Module 18 Key Words

Here's a list of key words for this module:

acid rain
agriculture
aquaculture
biodiversity
biological magnification
conservation
deforestation
desertification
ecosystem diversity
endangered species
extinction
genetic diversity
global warming
green revolution
habitat fragmentation
invasive species
monoculture
nonrenewable resource
ozone layer
pollutant
renewable resource
smog
soil erosion
species diversity
sustainable development

Module 18 Activity

TOPICS IN ECOLOGY

There are many critical ecological issues facing our society today. We are bombarded with so much that it's hard to tell which issue is really critical or someone's political agenda.

This activity will give you a chance to present your side of an issue of your choice with the goal of actually making a presentation to a local community group.

Listed below are some examples of possible topics. Try to focus on one that is of concern to your local community and design your presentation to address those concerns.

TOPICS
Acid Rain
History of Environmentalism
Air Pollution
Pesticides
Climate Change
Human Population Growth
Endangered Species
Rainforest Destruction
Water/Sewage Treatment
Recycling Programs
Conservationism
Water Pollution
Hazardous Waste Disposal
Solid Waste Disposal

PRESENTATIONS
The most popular method today of making a presentation is through slides. Using Microsoft PowerPoint or Apple Keynote is the best way to prepare the slides and deliver your presentation. A good presentation must have elements to it

that will keep the audience's attention. Your audience must leave knowing something they didn't when they sat down.

How to put together a PowerPoint or Keynote presentation is beyond the scope of this book. Suffice it to say, there are many good guides on the Internet that can take you through this process, as well as guides on how to present.

Pick a topic that you're passionate about, research it to the point where you're an "expert" on it and enjoy sharing your passion with your community or group.

USEFUL BIOLOGY WEB SITES

The following web sites represent twenty-four web sites that are general in their coverage. I tried to choose those web sites that will have appeal to anyone who is studying, or is interested in, biology. And some of them are just plain cool!

Biomes
The World's Biomes:
http://www.ucmp.berkeley.edu/glossary/gloss5/biome/

Cells
How Cells Work:
http://science.howstuffworks.com/cellular-microscopic-biology/cell.htm

Chemistry
Chemicool Periodic Table:
http://www.chemicool.com/

Classification
Animal Diversity Web:
http://animaldiversity.ummz.umich.edu/site/index.html

DNA
DNA from the Beginning:
http://www.dnaftb.org/dnaftb/

Human Genome Landmarks Poster:
http://www.ornl.gov/sci/techresources/Human_Genome/posters/chromosome/chromo01.shtml

Ecology
AE Mystery Spot:
http://www.accessexcellence.org/AE/mspot/

History of Ecology:
http://en.wikipedia.org/wiki/History_of_ecology

Evolution
AboutDarwin.com:
http://www.aboutdarwin.com/

Genetics
Probability of Inheritance:
http://anthro.palomar.edu/mendel/mendel_2.htm

History of Biology
History of Biology:
http://en.wikipedia.org/wiki/History_of_biology

Human Genome
Ethical, Legal, & Social Issues:
http://www.ornl.gov/sci/techresources/Human_Genome/elsi/elsi.shtml

National Human Genome Research Institute:
http://genome.gov/

Metric System
Brief History of SI:
http://physics.nist.gov/cuu/Units/history.html

Metric Conversion Table:
http://www.convert-me.com/en/Microscope

How Light Microscopes Work:
http://science.howstuffworks.com/light-microscope.htm

Population
Human Populations:
http://www.globalchange.umich.edu/globalchange2/current/lect
ures/human_pop/human_pop.html

How Many People Have Ever Lived on Earth?:
http://www.prb.org/Articles/2002/HowManyPeopleHaveEverLiv
edonEarth.aspx?p=1

Scientific Methods
Dispelling Some Common Myths about Science:
http://dharma-haven.org/science/dispelling-myth-magical-science.htm

Introduction to the Scientific Method:
http://teacher.nsrl.rochester.edu/phy_labs/AppendixE/AppendixE.html

Science Hobbyist Misconceptions Page:
http://www.amasci.com/miscon/myths10.html

Virtual Dissection
Edheads Activities:
http://www.edheads.org/

Virtual Eye Dissection:
http://www.eschoolonline.com/company/examples/eye/eyedissect.html

Virtual Lab
HHMI's BioInteractive Labs:
http://www.hhmi.org/biointeractive/vlabs/

These web sites I've used over the years, so all of the links worked at the time this page was posted. But the very nature of the web means that some of these could stop working at any time. If you find that any of them stop working, please email me so I can update this list.

Extending This Course

I offer a free newsletter full of tips on homeschooling science, and have both a weblog with science articles, links, recommended resources, and an online forum.
http://sciencelessonsforkids.com/sign-up

I am also constantly developing new course materials that integrate with the Week-by-Week Basic Biology Plan, as well as with my upcoming Advanced Biology: Homeschooling Week-by-Week, Earth Science: Homeschooling Week-by-Week, and Anatomy and Physiology: Homeschooling Week-by-Week.

You can find these modules linked from
http://sciencelessonsforkids.com/lessons